Life in ...

A Pond in
a Meadow

Distributed in the United States by
Smart Apple Media
1980 Lookout Drive
North Mankato, MN 56003

Text copyright © Sally Morgan 2000
Illustrations by James Field

ISBN: 1-929298-86-2

Printed in the USA

9 8 7 6 5 4 3 2 1

Library of Congress
Cataloging-in-Publication Data

Morgan, Sally.
    A pond in a meadow / by Sally Morgan.
      p.   cm. -- (Life in--)
    Summary: Describes the variety of plant
and animal life supported by the pond habitat,
explaining how these habitats are threatened
and how they can be preserved for the future.
    ISBN 1-929298-86-2
    1. Pond ecology--Juvenile literature. [1. Pond
ecology. 2. Ecology.] I. Title.

QH541.5.P63 M67 2000
577.63'6--dc21   00-022755
      4759

Editor: Russell McLean
Designer: Louise Morley
Picture researcher: Sally Morgan
Educational consultant: Emma Harvey

Picture acknowledgements:
Derrick Beavis/Ecoscene: 27t. Frank Blackburn/
Ecoscene: front cover br, 17, 22b, 27 Andrew
Brown/Ecoscene: 7t. Joel Creed/Ecoscene: 10-11.
Chris Gill/Ecoscene: 17t. Chinch Gryniewicz/
Ecoscene: front cover bl, 14tr, 25t. Jean Hall/
Sylvia Cordaiy Photo Library: 19t. Angela
Hampton/Ecoscene: 23b. Nick Hawkes/
Ecoscene: 24b, 24-25, 28-29. Papilio: front
cover tl & cl, 3, 7c, 11tr, 11tl, 12b, 13, 15l, 15br,
18, 19c, 19b, 20b, 20t, 21t, 22t, 23, 26t, 28b,
29b. Ken Preston-Mafham/Premaphotos: front
& back cover background, 6, 7b, 14o, 26b.
Barrie Watts: 21b. Robin Williams/Ecoscene:
11b, 12t, 16b, 16t.

Words in **bold** are explained in the glossary
on page 30.

Life in ...

A Pond in
a Meadow

Sally Morgan

Thameside Press

Contents

What is a pond?

A pond is a pool of water. The largest ponds are many yards across. The smallest are tiny pools. A pond can be home to many different types of animals and plants. It's a miniature world.

A pond forms in a hollow in the ground. The hollow traps rainwater and stops it draining away. Within a few days, the first animals fly in from other ponds.

Some ponds survive for only a few weeks before they dry up and disappear. Other ponds survive for months, or even years.

Some animals and plants spend all their lives in the pond. They are called **aquatic**. This means they can live in water. Other animals are visitors to the pond.

▲ This pond has formed in a hole left in the ground by mining.

▼ Insects and birds can spot a pond as they fly overhead.

▲ Many of the animals that visit ponds live in the surrounding woods.

In and around the pond

A pond can be divided into three zones—the open water in the center, the muddy bottom, and the shallows at the edge.

The surface of the open water is covered by floating plants and large waterlily leaves. Insects and small birds run across the leaves. Beneath the leaves, it is shady. The water is full of **microscopic** plants and animals called **plankton**. The undersides of the leaves are covered in **algae** and tiny animals such as **hydra** and insect **larvae**.

The bottom of the pond is covered by mud. Some animals crawl over the mud, while others burrow through it, looking for food.

waterlily

moorhen

tadpole

perch

caddis fly
larva

The water at the edge of the pond is shallow. Plants here have their roots in the pond and their leaves above water. Frogs, toads, birds, and small **mammals** find food and shelter among the plants.

rushes and reeds

kingfisher

dragonfly

vole

pond snail

iris

beetle larva

water beetle

newt

stickleback

frog

9

Life at the surface

The sun warms the water at the top of the pond. There is plenty of light for plants. The roots of the waterlily are in the deepest part of the pond, but its large, flat leaves float on the surface.

▲ Water spiders feed on insects that fall into the pond.

▲ Water striders move quickly over the water.

Some animals live on the surface of the pond. Water striders, whirligig beetles and water spiders are so light that they can move gracefully across the surface of the water without sinking. Other insects come to the surface to breathe. They trap bubbles of air between the hairs on their bodies. They use this store of air to breathe while they are underwater.

The water is full of microscopic animals and plants called plankton. These are too small to be seen with the naked eye. Plankton is an important food source for the larger animals in the pond.

▼ This whirligig beetle has trapped a bubble of air.

Life at the bottom

The bottom of the pond is very muddy. The mud forms over many years, from dead leaves and the bodies of animals that sink to the bottom. In some ponds, the mud is smelly and deep.

Worms, snails, and insect **larvae** burrow through the mud, looking for food. Flatworms, caddis fly larvae, and dragonfly larvae crawl over the surface of the mud. Midge larvae spend the day at the bottom of the pond. At night, they swim to the surface. Here they can feed safely in the dark.

▲ A caddis fly larva protects its soft body in a case, which it mak from tiny twigs or grains of sand

The mud is full of microscopic **fungi** and **bacteria**. They break down all the dead plants and animals. This releases food, or **nutrients**, that the pond plants need to grow.

◀ Dragonfly larvae hunt for food at the bottom of the pond.

Ducks and swans hunt for food on the bottom of the pond. They swim with their heads below the surface and their tails sticking out of the water. In deep ponds, some ducks dive to the muddy bottom to hunt. Turtles spend the winter asleep in the mud. During the rest of the year, they hunt fish, frogs, worms, and snails.

▼ This red-eared turtle is climbing out of the pond to warm-up in the sun.

Pond plants

Not all plants can live in water. The plants in the pond are aquatic plants. They are specially **adapted** to life in water.

Aquatic plants have lots of air spaces inside their leaves. The air makes the leaves **buoyant**. This means they float near the surface of the pond.

▲ The flower of a waterlily is brightly colored to attract insects.

There is plenty of sunlight for the plants on the surface. They use the light energy to make food. This is called **photosynthesis**. Plants give out a gas called **oxygen** during photosynthesis. On a sunny day, you may see tiny bubbles of oxygen rising to the surface of the water.

◄ Water ferns float on the surface of the pond.

Tiny plants called **algae** live in the pond. These thread-like plants float at the surface. In summer, the algae often cover the entire surface of the pond with a green mat.

Pond plants have many ways of staying alive during the cold winter months. Floating plants sink to the warmer water at the bottom of the pond. Pond weeds and waterlilies store food in underground stems called **rhizomes**. They use the food to fuel growth in the spring.

◀ Cattails live at the edge of the pond. At the top of each stem is a cylinder of brown seeds. When the seeds blow away, the flower turns white and fluffy.

▲ This pond is overgrown with plants. The water surface is covered with algae.

Web of life

The plants and animals of the pond depend on each other for survival. Plants are food for the animals. Animals which feed on the plants are called **herbivores**. They are eaten by meat-eating animals, or **carnivores**.

▲ Sticklebacks catch insects. In turn, the sticklebacks are eaten by larger fish and birds.

Dragonfly larvae and beetle larvae are ferocious carnivores. They have strong jaws to grip their **prey**. Fish are larger hunters. Small fish such as sticklebacks feed on smaller insects. The largest hunter in the pond is the pike (see page 20)—a huge fish that eats smaller fish. There are hunting birds too, such as the heron and the kingfisher (right).

◀ Predaceous diving beetle larvae use their strong jaws to catch food.

16

Pond snails (above) are important because they feed on the pond weeds and algae floating near the surface of the water. If the algae is not eaten, it soon covers the pond.

◀ Kingfishers eat fish and insects. They dive into the water to catch their prey.

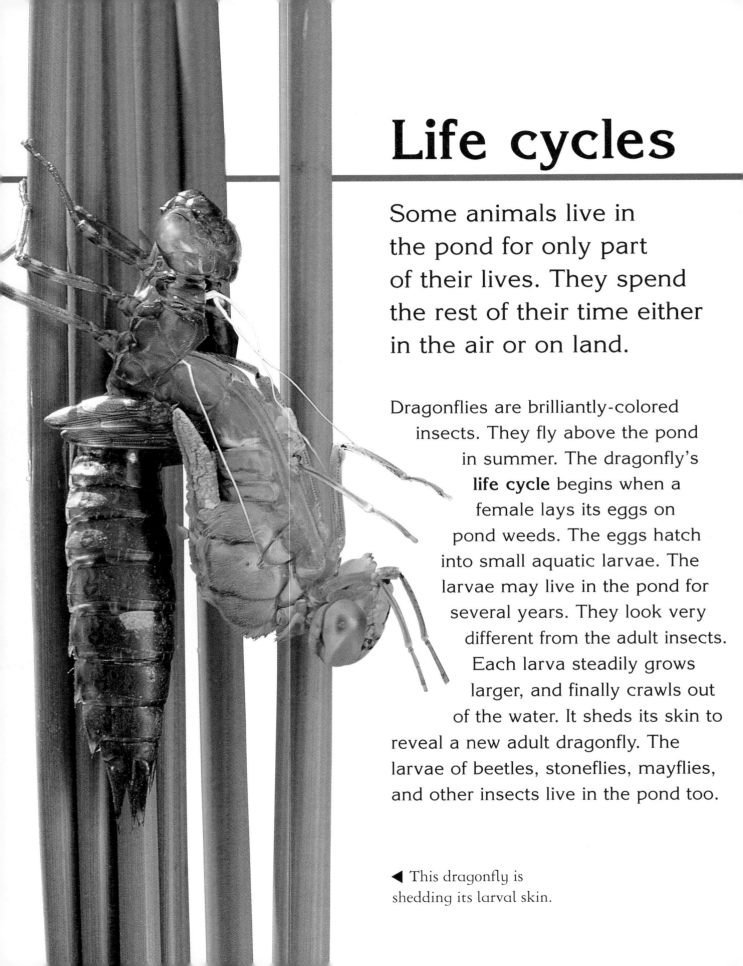

Life cycles

Some animals live in the pond for only part of their lives. They spend the rest of their time either in the air or on land.

Dragonflies are brilliantly-colored insects. They fly above the pond in summer. The dragonfly's **life cycle** begins when a female lays its eggs on pond weeds. The eggs hatch into small aquatic larvae. The larvae may live in the pond for several years. They look very different from the adult insects. Each larva steadily grows larger, and finally crawls out of the water. It sheds its skin to reveal a new adult dragonfly. The larvae of beetles, stoneflies, mayflies, and other insects live in the pond too.

◀ This dragonfly is shedding its larval skin.

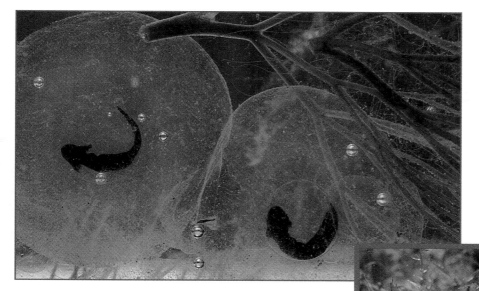

◀ Frogs lay their eggs in the water. The clumps of eggs are called frogspawn.

Frogs are well suited to living in the pond. They have long, powerful back legs and webbed feet to push them through the water. Frogs can live on land because they breathe air. Each spring, they return to the pond to **breed** and lay their eggs. The eggs hatch into tadpoles, which live in the water. When a tadpole has grown into a tiny frog, it is ready to leave the pond.

▲ Over a few months, a tadpole grows four legs and loses its tail.

◀ A female frog returns to the pond where it was born to lay its eggs.

Fish life

Fish are the largest animals in the pond. They range in size from small sticklebacks to the huge pike. Their slippery bodies move easily through the water. They use **gills** to breathe.

▲ Perch live at the edge of the pond. They eat insect larvae and small fish.

▲ A pike can open its jaws wide to swallow large fish.

Some of the fish are plant-eaters, or herbivores. Their gills **filter** plankton from the water. They eat algae too. Other fish, such as carp, feed on the dead bodies of other animals. The pike is a hunter. It lurks in the shadows, waiting for small fish or other animals to swim past. It has rows of sharp teeth to grip its prey. Very large pikes may feed on ducklings and cygnets (young swans).

▲ Tench live at the bottom of the pond.
In winter, they may bury themselves in the mud.

Young fish are called fry. They hatch from eggs, and grow quickly into adult fish. Sticklebacks are unusual because they lay their eggs in a nest. The male stickleback makes his nest on the bottom of the pond. Then he dances in front of a female, showing off his bright-red belly. If the female likes his nest, she will lay her eggs inside.

▶ In spring, a male stickleback turns bright red to attract a mate. For the rest of the year, the stickleback is a dull, browny-yellow color.

Bird life

Many birds build their nests around the pond. When their eggs hatch, the young birds are taken onto the pond for their first swim.

Ducks, coots, and moorhens paddle in the water, looking for food. Grebes and tufted ducks dive for fish. The largest birds are herons and swans.

▼ The purple gallinule walks about on floating plants, feeding on leaves and seeds.

The heron (right) is a hunter. It stands in the shallows at dawn and dusk, watching for fish to move. Some herons open one wing to cast a shadow on the water. Fish swim into the shadow, where they think it is safe. Then the heron stabs the fish with its long beak.

Mud and algae stick to the feet of birds. The mud may contain seeds, microscopic animals, and even fish eggs. The birds carry these plants and animals from one pond to another. In this way, each pond supports a wide variety of life.

◄ Warblers build their nests among the reeds. They feed on insects—especially dragonflies and midges.

▼ A heron has a long beak which is just the right shape for catching fish.

Swans spend most of the day on the water, feeding on plants, snails, small fish, tadpoles, and frogs. They build huge nests out of reeds and twigs. Swans **incubate** their eggs for 40 days. Then the cygnets hatch.

◄ Cygnet feathers change from gray to white at the end of the summer.

Drying out

The pond is always changing. Each year, the layer of mud at the bottom grows thicker and the pond becomes smaller.

The plants in the pond grow larger over time. Their dead leaves sink to the bottom, adding to the mud. During hot weather, the pond shrinks as the shallows dry out. The plants grow out into the deeper water in the middle of the pond. Eventually, even the deepest areas may become filled with mud. Soon, the pond dries out in summer. It becomes a swamp or **marsh**. Then trees and **shrubs** move in. Within a few years, the pond disappears completely. The area becomes a wet woodland.

▶ As the pond becomes shallower, the plants around the edge grow in toward the middle.

The pond can be saved if the thick mud is removed, or **dredged**. Some of the plants at the edge may have to be pulled out also. This lets more water into the pond, so there is more room for the aquatic plants and animals.

◀ These volunteers are clearing mud and reeds from an old pond.

◀ This pond has almost disappeared. Plants have grown into the middle and there is no open water.

Around the pond

A meadow surrounds the pond. Meadows are small fields filled with grasses and other flowering plants.

In early summer, grasses produce spikes of tiny green flowers. These flowers release huge amounts of **pollen** into the air. Among the grasses are colorful plants, such as ox-eye daisy, poppy, meadow cranesbill, and cornflower.

▼ Honey bees become covered in pollen when they visit flowers.

▲ Butterflies drink the sweet nectar made by the flowers in a meadow.

During the summer months, the meadow is alive with bees, butterflies, and other flying insects. The flowers are brightly colored to attract them. The insects visit the flowers to drink their sugary **nectar**. At the same time, they pick up pollen which they carry to another plant. This is known as **pollination**. Once a plant has been pollinated, it can produce seeds.

In late summer, the meadow grasses are cut to make hay (above). The hay is used to feed farm animals during the winter. If the meadow is not cut, shrubs and trees will start to grow in it. The meadow will turn into a woodland.

▼ Poppies, cornflowers, and ox-eye daisies are some of the flowers growing in this meadow.

Looking after ponds

If a pond is allowed to fill with mud, it will dry up. Meadows can become overgrown with trees and shrubs. We need to look after, or **conserve**, these **habitats**.

Sometimes, ponds are filled in and meadows are plowed up to make more land for crops, roads, or houses. **Pollution** damages ponds also. Farmers use chemicals, such as **pesticides** and **fertilizers** on their fields. These may drain into the pond and kill fish and other aquatic animals. Some ponds are polluted by **sewage**.

▲ Fertilizers cause algae to grow very fast. A thick mat, called an algal bloom, covers the surface of the pond.

◄ Garbage is often dumped in ponds, where it can harm the pond animals.

Many people dig a pond in their backyard. The first insects arrive after a few weeks. Even the smallest pond can be home to many aquatic animals, including frogs. All over the world, frogs are dying out because they are so sensitive to pollution. Backyard ponds are helping to save them.

◄ A pond is a good place to learn about wildlife.

Glossary

adapted Made suitable.

algae Tiny green plants that live in water. They do not have leaves, roots, or flowers.

aquatic Living in water.

bacteria Tiny, single-celled organisms which are too small to be seen with the naked eye.

breed To make babies.

buoyant Able to float.

carnivore An animal which eats other animals.

conserve To protect habitats and wildlife.

dredge To remove mud from the bottom of a pond, lake, or river.

fertilizer Nutrients which help plants to grow.

filter To remove solid particles (like food) from a liquid (like water).

fungus (plural **fungi**) An organism that is neither animal nor plant. Most fungi are made up of tiny threads that grow through the soil.

gill An opening on the side of a fish's head, through which it breathes.

habitat The place where a plant or animal lives. Habitats include forests, ponds, and seas.

herbivore An animal which eats only plants.

hydra A small organism that lives in freshwater. It has a soft, tube-like body, and a mouth surrounded by tentacles. The bottom of the tube is attached to a rock.

incubate To keep warm.

larva (plural **larvae**) A young insect that has hatched from an egg. The larva eventually turns into an adult.

life cycle The changes an animal or plant goes through, from birth to death.

mammal A type of animal that feeds its young on milk from the mother's body.

marsh An area of soft, wet ground, often near a lake or river. Another name for a marsh is swamp.

microscopic Too small to be seen with the naked eye.

nectar Sweet, sticky liquid made by flowers to attract insects.

nutrients Chemicals that plants and animals need for healthy growth.

oxygen A gas in the air that most plants and animals need to survive.

pesticide A chemical that kills insect pests, such as greenfly and locusts.

photosynthesis The way green plants make their own food using sunlight. The leaves use light energy to combine carbon dioxide and water to make sugar and oxygen. Plants use the sugar as fuel.

plankton Tiny plants and animals that live in water. They are food for fish and other aquatic animals.

pollen Yellow, powdery grains made by the stamens (the male parts) of flowers.

pollination Carrying pollen from one plant to another, so that a flower can make seeds.

pollution The release of harmful substances into the environment. Pollution can damage or poison living things.

prey Animals which are killed by other animals for food.

rhizome An underground stem.

sewage Garbage and human wastes, carried by drains and sewers.

shrub A small, tree-like bush.

OTHER READING

The following titles give more information about the plants and animals in this book.

In Fields and Meadows (*Animal Trackers* series),
Tessa Paul, Crabtree, 1997.
(ISBN: 0 865 05593 9)
A Freshwater Pond, Adam Hibbert, Crabtree, 1999.
(ISBN: 0 778 70147 6)
Look Closer: River Life, Frank Greenaway,
Dorling Kindersley, 1998.
(ISBN: 0 789 43478 4)
Rivers and Lakes (*Wonders of Our World, No. 1*),
Neil Morris, Crabtree, 1998.
(ISBN: 0 865 05846 6)
Rivers and Streams (*Exploring Ecosystems* series),
Patricia A. Fink Martin, Franklin Watts, 1999.
(ISBN: 0 531 15969 8)
The Young Oxford Book of Ecology, Michael Scott,
Oxford University Press, 1998.
(ISBN: 0 195 21428 5)

Index